Recirculation Indoor Shrimp Farming

Recirculation Indoor Shrimp Farming

ZEAH Zero Exchange Aerobic Hetatrophic Water

Karlanea Brown

Niche Pressworks
Indianapolis

RECIRCULATION INDOOR SHRIMP FARMING

ISBN-13: 978-1-946533-31-9

Published by Niche Pressworks
http://NichePressworks.com

Contents

Introduction 1

Expect the Unexpected 5

- So, You Married a Farmer? 5
- Honey, I Think You May Have Hit Something 7
- A New Home 8
- A New Plan 9
- Did Someone Say Shrimp? 10

The Basics 13

It All Started With a Little Multitasking 15

Amazing Little Creatures 17

- What Do You Mean, They Jump? 17
- Our Little Divas 18

The Whole System 19

Saving Our Oceans 27

Bumps Along The Way 29

A New Way of Doing Things 33

- The Baby Shrimp 33
- The Intermediate System 34
- The Volcano Effect 35
- Heating and Facility Arrangement 36
- Drainage 37
- Power Usage 38

The Standards We Set 39

So, You Want to be a Shrimp Farmer? 41

Regulations and Certifications 47

- Regulations 47
- Certifications 48

Tours and Community Involvement 51

Do You STILL Want To Be A Shrimp Farmer? 53

Let's Get Shrimping! 57

Testimonials 59

About the Author 61

Introduction

If anyone, and I mean ANYONE, had told me 20 years ago that today, I would be running one of the most successful saltwater shrimp farms in the United States, I would have told them they were crazy! Really, really, crazy. Add to it that I would be doing it from my rural Indiana home, and I probably would have run the other way screaming!

But honestly, I wish we had thought of it sooner, so we could enjoy it much longer.

Shrimp farming is not something I ever wanted to do. It's not something I ever even thought of, much less thought of doing. And, it certainly isn't something I ever thought I would enjoy. Am I making my point here? The truth is, shrimp farming is something I have come to love. It is truly my passion, work I love doing, and I enjoy every single day of it.

My alarm goes off at 5am every morning. I get up, walk the dogs, eat breakfast, quickly catch up on the news, and am usually answering customer emails before 6am.

I'm in the office by 7am and by 8am I have already walked the farm, checked the tanks, turned the lights on, picked up and disposed of any "suiciders" (shrimp who managed to jump out of the tanks overnight), and started my daily water testing on the tanks.

All of this before I can even start feeding in the nursery. You see—I need the water testing to be done before I can figure out how much to feed the baby shrimp. And—they have to be fed about five times a day. After all, we have to take care of our babies!

Around 9:30am, after I've cleaned everything—tanks, glass, air lines, floor, and anything else that needs it—I start back up with emails. Did I mention that we get a lot of emails? Oh—and we open to the public at 9am, so I also talk to anyone who pops in, answer the phone, give tours (yes, we offer a short, 20-minute tour to the public), and basically handle the front-end side of the business.

No two days are alike, but I will tell you—I am never, ever bored.

On the flip side of the business, we are on the road roughly two weeks out of the month, delivering post larval shrimp (PLs for short) or helping new farmers get set up.

We sell several customizable "Start-up Packages" which may include setup of the equipment. We have set up 42 shrimp farms in the United States, and as of the printing of this book, are in the process of setting up five more, which are located in Switzerland, France, Ukraine, Canada and Peru. While we're on the road, we have some wonderful employees who do the feeding and testing for us. They also harvest shrimp or crawfish for customers, move the shrimp to new tanks as needed, and clean throughout the day.

We have 26 shrimp production tanks and 10 nursery tanks. We also have 24 crawfish tanks, two tilapia tanks, and an indoor salt water aquaponics system in which we are growing mangrove trees, but I'll save that for another book!

We close the front office at 4pm but are hardly done at that point. We still have to catch up on emails (again), close the books, do more feeding (did I mention the nursery shrimp have to be fed five times a day?), and close up the farm for the night. If I have any free time, I usually do some research. There is so much research to stay on top of in this industry. Plus, I've always wanted to get into oyster farming …

I'm usually in bed by 8:30 or 9:00pm so that I can get up and do it all over again.

All of this from a woman who at twelve years old, wanted nothing more than to be the next, big fashion designer. I had never even been on a farm, let alone knew what a pig or cow looked like, other than from a picture.

What we do here on our farm is generally called aquaculture, which is the farming of fish, crustaceans (like shrimp), mollusks, aquatic plants, algae and other aquatic organisms.

The funny thing is that I had never even heard the word aquaculture until my husband and father-in-law decided they were going to start raising tilapia.

But I'm getting ahead of myself here. Let me tell you how all of this got started. There have been laughs, tears, and a lot of surprises over the years. And I wouldn't change a single thing. I love doing this now, and when I am 70 years old, I will still love it. In fact, I plan on being 90 years old, still getting up at 5am, and still working on the farm every day.

So, let's start at the very beginning …

Chapter 1

Expect the Unexpected

I never imagined this kind of life for myself. Just the opposite, in fact. When I was 12 years old, all I wanted was to be a fashion designer. I am not talking about a no-name, run-of-the-mill fashion designer, creating T-shirts and jeans for the likes of Walmart.

I was going to be the next Coco Chanel or Gianni Versace. My designs were going to be seen on runways around the world and coveted by women and men alike! I even went to college for fashion design and had every intention of living in New York City and fulfilling my dreams!

As you've probably already figured out, that didn't happen.

So, You Married a Farmer?

Everything changed in 1984, when I met my future husband, Darryl. I don't mean that meeting him changed my mind; I truly thought that I was going to be able to convince him to move the family farm to New York City, so he could farm, and I could be the next big fashion designer.

As you can see, I didn't make it. I stayed in a small town in Indiana and married a farmer, as was his father before him, and his father's father before him. The funniest part is that I was raised in Lafayette, Indiana. I didn't even know what a cow, a pig, or a chicken looked like before I met Darryl. Yet, I married a hog farmer.

Oh sure, I had seen pictures of animals in books and magazines, but I had absolutely no reason to go to a farm. I knew my bacon came from a pig, my steaks came from a cow, and my chicken legs and wings came from chickens. And that was the extent of my livestock knowledge.

Now that I think about it, I probably didn't even know the term "livestock" back then. I enjoyed cooking steaks and bacon, but as far as I knew, they came from the meat section at the grocery store. I didn't need to know any more than that.

Since I couldn't move to New York City and become the fashion designer I knew I could have been, I did the next-best thing: I became an interior designer. There's not much of a call for fashion designers in rural Indiana. My husband Darryl happened to run his own home building company in addition to working the family farm, so that was actually a pretty natural fit for my interior design business.

I worked in interior design for about 10 years, and I was really good at it. Then my son was born, and I chose to stay at home with him until he went to school. After 10 years of marriage to Darryl, I still didn't have anything to do with the farm aside from handling some menial chores. That's because I was pretty much banned from the farm during our first year of marriage.

Honey, I Think You May Have Hit Something

Darryl's family, the Browns, had plenty of reasons to ban me. I mean, when you wipe out five miles worth of mailboxes driving a tractor down the road and have no idea you hit even a single one, or when you run a "honey" wagon through the back of a hog barn, resulting in 100 pigs running wildly around the yard (did I mention that pigs are fast and nearly impossible to tackle?) with my father-in-law laughing so hard he couldn't help me catch them, you probably deserve to be banned!

It's possible that the family may have had other good reasons to keep me from helping on the farm. I still think to this day that it might have been partly Darryl's fault as well; I mean, it can't be all my fault, can it?

After my son went to school, I didn't want the crazy hours of interior designing anymore, so I went to work for the school he attended. What better place to work and have the same hours as my son? This way I wouldn't miss any ball games or programs he might have been in. I worked in remediation and then managed the school cafeteria. As you can see, I was still gaining zero farming experience.

My husband, on the other hand, has a degree in agriculture from Purdue University and grew up raising pigs, turkeys, cows, corn, and soy beans. His family ran what most people in the '80s thought of when they thought of a farm: all kinds of different livestock running around. Kind of like in Charlotte's Web, without all the talking. Pigs were the main source of live-stock income for the Browns.

At our first place, we had 5000 head of hogs on the property. I actually really liked working with the pigs; it was my relaxing period of the day. Cleaning stalls (one of the few things I wasn't banned from!) was therapeutic for me. Maybe it was because it was quiet, or perhaps because it was simple, menial work.

Even though it usually smelled pretty awful in the stalls, after the stress of dealing with people agonizing over the color of their drapes, and whether or not they matched the hideous wallpaper they had picked out (until I changed their mind on the wallpaper), I really enjoyed it.

A New Home

After we'd been at this location for about five years, the neighboring farmers began selling their property to development companies for future subdivisions. We knew we were going to be the most hated neighbor in the area. People like the country; they just don't like the smell or the sound of a hog farm. Have you ever heard a hog farm?

So we, along with my in-laws, Russell and Marie Brown, moved to where we are now—Fowler, Indiana, to set up a new hog operation. Fowler is the county seat of Benton County—located in the northwest part of the state, along the border with Illinois. It's a quaint farming community with a population of only 8000 people.

To give you an idea of just how rural we are, we live on a stretch of road where for four miles in either direction, there are only eight houses. In the summer, it's rows of corn or soybeans for as far as the eye can see. There might be a wind turbine or two in your vision—Benton County was the first county in the state of Indiana to have turbines, and we now have over 400 of them and counting!— but mostly it's just corn and soybeans.

Benton County is now also known for our shrimp farm, which is something you don't hear very often. Cows and pigs, maybe. Shrimp? Never!

In 1992, shortly after we broke ground on the new hog buildings, disaster struck. Hog prices dropped to an all-time low. It was so bad that we would have lost everything if we had continued in hog farming.

A New Plan

Now we had to switch gears, and quickly. When I say "we," I actually mean "they." I was banned from doing any work on the farm, remember? I was allowed to help out only when absolutely necessary, and believe me, I was their last choice. I was allowed to bring dinner when they were out in the fields, and to clean stalls, but that was about it.

Now, Darryl and his dad had to figure out what they were going to do. Being very traditional farmers like they were, they felt a farm should have livestock. They couldn't continue to raise hogs, going back to cattle was too much work, and Darryl couldn't bear to raise chickens. He had worked on a chicken farm in high school and said he would never own chickens. This was when they came across aquaculture and tilapia farming. This intrigued them both.

For the next 15 years, they continued corn and soybean farming while they did research on aquaculture. They visited Purdue University's aquaculture division, and asked what had to have been a million questions about how to raise tilapia. There are so many ways to raise so many different types of fish (or shrimp for that matter, but we will get into shrimp later).

They really wanted to raise tilapia. Tilapia is a hardy fish, so you can make some mistakes along the way and they will survive. One of the biggest bonuses with farming tilapia is the fact that they only take six months to raise, unlike some other fish that can take up to seven years. Tilapia offered faster turnover, and a seemingly better way to make money.

The Browns were so set on raising tilapia they actually went out and bought ten 15,000-gallon raceway tanks to get things going. For those of you who don't know what a raceway tank is (I know I didn't at the time), it is a long, rectangular tank, usually made of fiberglass. These tanks will hold 6,000 to 15,000 gallons of water and are used for flow-through aquaculture.

After years of research, they finally visited a farm that was raising tilapia and was selling a tilapia system to get them started. They went, and I stayed behind. You have to remember: I didn't, and wasn't ever going to have much to do with this fish farming. I was not interested in it at all.

Did Someone Say Shrimp?

It was during this visit that they discovered the tilapia farm had just developed a shrimp pump system, and they told Darryl and Russell all about shrimp farming. Shrimp farming? *What happened to tilapia?* This was apparently even more interesting than the tilapia.

They came home very excited and were suddenly talking all about shrimp. The cost to get into the shrimp business was half the cost of getting into tilapia. So for the Browns, this was a good thing to try. *Okay ...*

You have to realize when someone talks about the same thing for 15 years you tend to tune them out and nod your head as if you are listening. You're just hoping you are not agreeing to something that could get you into trouble later down the line.

Every time Darryl mentioned shrimp, my eyes got really big, and I know my ears perked up to listen to the details of the conversation. I was like, *we can actually raise shrimp indoors?* I like shrimp more than tilapia, and a shrimp is so much easier to clean than a fish, right? Why not raise shrimp? My concept of shrimp farming at that point was that I would go out to the barn and dip out my shrimp and cook them for dinner. I also thought there were hundreds of shrimp farms across the United States. Boy, was I wrong, on pretty much all counts. ...

The family agreed that our new "livestock" was going to be shrimp. We decided on the type of building we would build. A building with enough room for six 18-foot tanks was about all the space we could spare, and we knew we would fill it. The new building would be squeezed between the greenhouses on one side, where we raised annuals and perennials that we sold at the farmers' market, and the building that housed our combine. On the back side, we were only about five feet from the cliff that dropped down into the creek. This was about the time I found out that we were going to be only the third privately-owned shrimp farm in the United States. *Ummmm ... okay, let's do this!*

So, we celebrated as we broke ground for the "Brown Family Shrimp Farm." It would later be officially named RDM Aquaculture. The letters stand for Russell, Darryl, and Marie.

Three weeks later, my father-in-law Russell passed away suddenly. After all the years of talking about this, planning it out, doing research, and preparing for it, it was finally happening. And we lost Russell, just like that.

What we didn't know was that Russell had been ill for some time; he just didn't tell anyone about it. He never let on to any of us that anything was wrong. We had no idea, and now he was gone.

This put my husband in a very difficult situation. He had to decide if he was going to continue in the venture he started with his father, or if he was going to focus primarily on his building company. I told him I would support him no matter his decision, and I would help him if he chose to continue with aquaculture.

Honestly, I lied to him. I had absolutely no intention of helping, unless he really needed me to. I didn't want anything to do with this, other than dipping out my dinner. The shrimp farming was his thing. I had my own things going on at that time. Besides, I figured his mom would help him.

After much deliberation and debate, Darryl chose to continue with the shrimp venture he and his dad had started. His body wasn't holding up well doing home construction. His back hurt all the time, and he just didn't enjoy it anymore. I also think it had something to do with continuing his father's legacy but he never actually mentioned that. Being the good wife that I am, I supported him in his shrimp venture (from a distance, of course).

It took us one year to complete the building and about three months to get all of the equipment up and running. So, in July of 2010, we were officially in the shrimp business and received our very first shipment of post larval shrimp (PLs for short). *Now the FUN begins!* I had no idea just how true that was.

Chapter 2

The Basics

We raise our shrimp in huge round plastic tanks using a saltwater zero-exchange aerobic heterotrophic (ZEAH) system. That's a very big, fancy term, but let's break it down. "Zero-exchange" means we still have the same water in our tanks that we started with years ago. Rather than pumping all the water out of our tanks when it gets dirty, we only have to replace the water that is lost to evaporation. That's because the specialized bacteria in our water keep it clean. Our system does the same thing as the ocean by cleaning itself up. The ocean has no filter system other than bacteria and aquatic life to clean her up.

"Aerobic" refers to a system that requires oxygen. Our shrimp would die if the oxygen levels got too low, so we constantly have to keep the water circulating to keep it aerated.

"Heterotrophic" refers to the bacteria that grow in our tanks. Heterotrophic organisms can't make their own food, so they have to eat, just like you and me. The bacteria attach to the bits of uneaten food and the shrimp waste floating in the water, forming clumps called "biofloc".

Shrimp waste is mostly ammonia. When ammonia levels in the water get too high, it kills the shrimp. Some of the bacteria in the biofloc "eat" by turning the ammonia into nitrite, which is less toxic than ammonia, but still not great for shrimp. Thank goodness we have still other kinds of bacteria in the biofloc. These "good guy" bacteria turn the nitrite into nitrate, which won't hurt the shrimp. The biofloc also serves as a food source for the shrimp, which helps keep our feed costs down.

Heterotrophic systems came into large-scale use about 30 years ago in Belize, primarily for shrimp and fish farming.

I'll talk more about why all this is important in later chapters. You'll see that shrimp farming is all about paying really close attention to water quality!

Chapter 3

It All Started With a Little Multi-tasking

I just realized that I completely forgot to tell you how *I* got involved in our shrimp business. Before we got our first batch of PLs, Darryl played with the water test kit we got, and it took him five hours to test one, single tank. *Uh-oh.* Let's do the math on that one. We were going to farm six tanks, and he still owned his building company. If it took him five hours to test the water in one tank, it was going to take 30 hours to test all the tanks. The tanks need to be tested daily, and last time I checked, there were still only 24 hours in a day. You get the picture.

I multitask everything, so I figured I would simply show Darryl how to multitask his tank testing. Remember, my degree is in fashion design, not chemistry, biology, or farming. I know about fabric and design. I know about paint and wallpaper. I know about making my customers happy and their homes beautiful. I'm also a mother. I do multitasking very, very well.

I came out on day two to show him how to do the testing in a faster fashion, and I never left. That was eight years ago. Who knew I would love the chemistry side of this business as much as I do? Who knew I would be good at it?

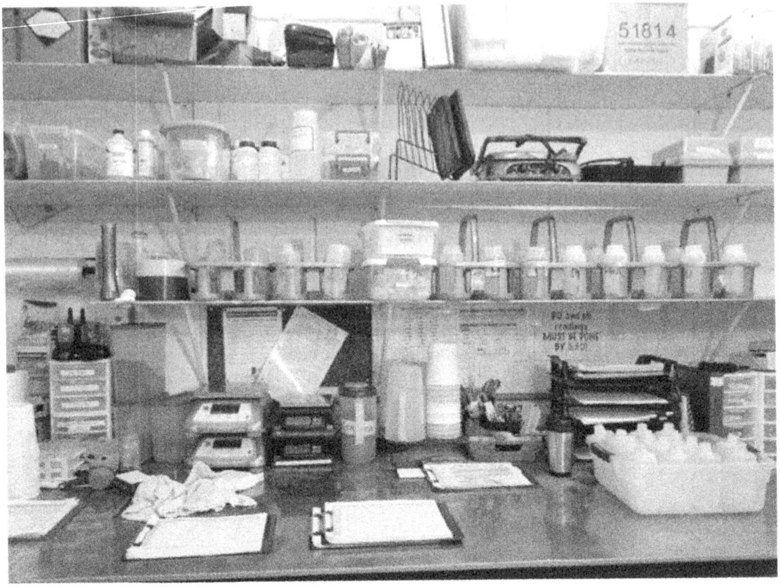

In all the time we have been up and running, I have missed only three days. Two of them were because I lost my voice talking about shrimp farming, and one was a day I had to pick my dog up at the airport. Even when we are helping set up new farms or delivering PLs to farmers, I am still in contact with the people back at the farm. Never in a million years would I have guessed this would become my passion and my love.

Even though I had never even heard the word aquaculture until we started the shrimp business, I simply cannot see myself doing anything else. This world is so intriguing. In aquaculture, there are so many species of creatures which you can grow indoors. Why not try them all? I'd recommend starting with one, mastering it, and then going on to the next one. The possibilities are endless!

Chapter 4

Amazing Little Creatures

When we first got started raising shrimp, I knew very little about them. I didn't even know they had heads. I mean, you know they have to have heads, right? The reason I had never seen a shrimp with its head still attached is that you cannot freeze shrimp with the heads still on. Since I had only ever eaten shrimp that had at one point been frozen, I'd never seen the heads.

What Do You Mean, They Jump?

After about two months working with the shrimp, I accidentally learned about the heads when I had one jump out of the tank and land in my hair. I had long hair, pulled back into a ponytail, and somehow, one little jumper had landed in my hair. I screamed like I had never screamed before. Who knew shrimp could jump out of their tanks?

That was IT! I was never again setting foot in the barn. As far as I was concerned, they were mean, aggressive little creatures. But I knew I was going to have to go back in the barn at some point. Heaven knows, Darryl could not test the tanks!

Over the years, I have also had shrimp land down my shirt. I now wear two shirts. I even cut all my hair off so those little jumpers couldn't land in my hair and grab hold of my pony tail ever again.

These days I like the shrimp. I think they are beautiful little creatures of God. They are very smart, very social, and are very interesting to watch.

In the beginning, when we were using one-horsepower pumps, the shrimp would come to the top of the water during feeding time. I could drop one pellet at a time, and they would grab it and float on their backs like sea otters and eat off their bellies. Then they would swim against the current and get back in line for more food. Now you can understand why I fell in love with these little guys. They have great personalities. Some people think they are ugly, but I think they look like mini dragons with a very simple approach to life. They just chill and clean the bottom of the ocean … or in my case the tanks.

Our Little Divas

At RDM Aquaculture, we raise *Litopenaeus vannamei*, commonly known as Pacific white (or white leg) shrimp. When we were getting things set up and trying to decide which type of shrimp to raise, we were told these are the hardiest, and decided they were the best choice for us since we were just starting out. When you're doing this indoors you really need a resilient creature to raise.

I soon learned that if *L. vannamei* is considered the hardiest species, I don't think I could handle any other type of shrimp. We call them "divas," and with good reason. They are very temperamental, and don't like change of any kind. The slightest change in their environment can kill them. For example, if the water temperature decreases a mere two degrees over the course of a couple of hours, you can actually lose half a tank of shrimp. If the two degree temperature drop is spread over 24 hours, you might lose only 20-30 percent of your shrimp (the true divas).

Chapter 5

The Whole System

There is not a lot of information out there about how to raise shrimp, but as you can imagine, one of the most important factors in a shrimp's environment is the water. The success of the whole shrimping system rests on water quality. It's like the air you and I breathe. Too little oxygen and we get confused, restless, and suffer from severe headaches. Too much oxygen causes seizures, dizziness, and sometimes loss of consciousness. And we're not even that delicate!

Water Quality

In many ways, we are "Guardians of Water" rather than shrimp farmers. In shrimp farming, it is all about the water. We are trying to create and maintain a "Garden of Eden" for the shrimp. These little guys need the perfect environment in order to thrive and grow; and we work very, very hard to make sure they have it.

We test our tank water once each day for nine different factors: dissolved oxygen, nitrites, temperature, salinity, alkalinity, ammonia, CO_2, pH, and settled solids.

We feel that if we know what is going on with the water we can maintain or change it with little or no effect on the shrimp. Nothing happens overnight. It takes at least four days before neglect will show up in your tanks. The outcome of neglect is pretty obvious: shrimp start dying.

We made a lot of mistakes in the beginning and we still make mistakes today, but we are learning from them. One of the biggest mistakes we made was with our biofloc.

We've already talked a little bit about the biofloc, but let's go into more detail. At its core, the biofloc is a waste treatment system. In our case, the biofloc replaces the natural organisms that wild shrimp are exposed to with beneficial bacteria and other microscopic life. Biofloc is also a feed source for the shrimp.

When we started this using this type of system we were told to keep our biofloc levels at around 60mg/l, and we did. It took us about one year to get to that level. We had a 30-40% shrimp survival rate at that time. We were stocking 16000 to 20000 post larvals in each of our production tanks per month, and getting only about 7000 of them to survive. That translated to only about 250 pounds of 20-count shrimp.

In order to maintain the right biofloc levels, we use what are called settling tanks or floc tanks. For my purpose, I will call them settling tanks. A settling tank is a 300-gallon portable tank that sits outside of each shrimp tank. An air lift pump utilizing an air stone and gravity moves the water from the shrimp tank into the settling tank, and the biofloc comes along with it. The biofloc falls to bottom of the settling tank, and the clean water is returned to the shrimp tank.

The pump is only turned on when the test result tells us that biofloc levels in the shrimp tanks are too high, and the test will also tell us when the levels have returned to normal. We call it the "Heart of the Operation," because it's so important to the shrimps' survival.

As I mentioned, when we first started we were told to keep our floc at about 60mg/l. But the shrimp kept hitting a wall. They became lethargic and didn't produce very well.

One day, I accidently forgot to shut off one of the pumps, and let it run way too long. When I tested the water in the shrimp tank, the biofloc levels were at 1 mg/l. *Um ... way low!* In panic mode, I realized that there was no way to get the bacteria back into the tank.

The only way to get redemption for my mistake was to hope the pump would break down. I was not about to tell my husband about the mistake I had made. I just kept hoping the pump would go down, the shrimp would all die, and I would be off the hook.

As luck would have it, the pump didn't go down and the shrimp actually grew thirty days faster than their counterparts in the other tanks! The best part was this was also the best production tank we had up to this point. We had produced over 200 pounds of shrimp out of this tank. The bonus was that they never hit the wall, never seemed lethargic, and they actually looked healthier than the shrimp in the other tanks.

Okay, so being the scientist I suddenly had become, I had to figure out why these shrimp were so much healthier than the rest. Was it the lower biofloc level? Was it some other mishap? Were they just healthier when they went in? I had to figure it out!

My husband was also trying to figure out why these particular shrimp grew faster and never hit the wall. I still didn't say anything, I just shrugged and said, "I have no idea." Yes, I am a coward, but if the beneficial effect didn't last, I didn't want to take the blame. Maybe I would fess up later.

So I started testing by myself by lowering the biofloc level on a second tank. The first tank still had low biofloc levels. Three months later, we had the same result out of the second tank. Not only more shrimp, but they grew 30 days faster, and looked so much healthier than before. *Now it was time to say something to Darryl.*

When I told him what I did, his response was, "Why would you decide to do something like this without talking to me first?" I told him, "It just came to me!" No, I didn't tell him I accidently left a pump on too long, and I didn't tell him I altered a second tank without talking to him. It doesn't matter, it worked!

To this day, I don't think he knows the truth about what I did. I know I haven't told him. *Hopefully, he doesn't read this book.*

We then lowered all the floc levels in the shrimp tanks and it made a big difference. We started getting better survival rates.

Population Density

Another change we made was to lower our stocking densities from 16,000 to 7000 per tank for our larger tanks (based on 25 shrimp to the square foot). This gave us a chance to improve the survival rates even more. We were told the more we put in the tanks, the more we would get out, but we learned that shrimp do not like to be crowded. They actually hang out at the bottom of the tank and not in the volume of the tank the way fish do. In my mind, if you overstock, you stress the shrimp. They will start to fight each other for space and for food. Simply put, if they get too stressed, they die.

This was probably the best and biggest "mistake" we have made since we've been shrimp farming. Who would have thought it would turn out to be such a positive mistake!

Life Cycle

When we get the PLs from the hatchery they are only 10 days old. The post larval shrimp spend 25 days in the nursery, followed by 25 days in the intermediate tanks, and then 90 days in a production tank. This makes for a total life cycle of 150 days. Each phase has its own protocol. There is no one-size-fits-all.

For example, the nursery division goes through five stages of feed. They get fed five times a day on a very strict schedule. This "baby food" contains 50% protein, and is specially made for shrimp. We buy our feed from Zeigler in Pennsylvania.

We start them on a liquid diet (we use a product called Artemia) and work our way up to a powder and then to tiny granules. Remember that PLs start out so small that if the food particles are too big, they can't eat them.

When they move into the intermediate tanks they start on what we call the "Gerber Graduates" feed, which consists of larger granules, and has 40% protein content. Once they go into the production tanks they are put on the "big" feed (35% protein), and stay on that until they are sold onto someone's plate.

Each division also has its own set of testing and heating protocols. Just like newborn babies, the PLs like warmth and don't mind being overcrowded. When they go into intermediate tanks, we lower their temperature by 2 degrees Celsius (saves on heat) but they are still overcrowded, and they don't seem to mind it. After all, they are only in the intermediate tanks for 25 days. It's kind of like when kids go to elementary school. They like having all their friends close by to play with them.

When the shrimp go into the production tanks, they are called juveniles and they no longer like to be overcrowded. Just like teenagers, they want their space and only their select few friends around. I know it's a simplistic analogy but this is how their life cycle works. I don't think they have cliques like we did in high school, but we know crowding stresses them out and a stressed out shrimp usually doesn't live very long.

Testing the Waters

We do testing every single day. I know this might sound extreme and expensive. It is an added cost of .02 cents per pound of shrimp to test every day, but when you do not know what is going on at the bottom of your tanks the water is going to tell you. You really need to know. Bad water quality doesn't happen overnight, but it can happen in as little as four days. Not knowing can end up being a tankful of dead shrimp.

I learned that lesson the hard way. One time, we didn't test for three days. The previous tests were identical and everything was looking perfect. So why test again? It takes so much time. We now look back on those days as the "arrogant stage." I was working in the school cafeteria at that time and honestly, I just wanted to get home.

When I walked into the barn on day four, we had two full tanks of dead shrimp. That was 400 pounds of shrimp … all dead. The pumps were still working so I couldn't blame it on them. I went to look at water quality results from the day before, but wait, I didn't have any because I didn't test the day before. So I did a full test on the tanks and the only problem was that the alkalinity levels had dropped significantly.

The first day I skipped the water test, I would have caught the lower alkalinity levels and added baking soda in time to correct the issue. Now we test every day just for that reason. No exceptions. No more taking chances.

When we test our tanks we are testing for dissolved oxygen, temperature, salinity, pH, ammonia, nitrite, alkalinity, CO2, and settled solids. Since our water is brown due to the bacteria and feed in suspension, we cannot see the bottom of the tanks, so we also do what we call "raking" the bottom of the tanks. This lets us see what is on the bottom of the tank along with any dead shrimp, and it lets us see what the shrimp look like. The shrimp tell us the story of what is going on and the testing just confirms the findings.

What we look for is long antennas—this means the shrimp are happy and healthy. We also look for war wounds or scabs on their bodies.

One or two shrimp with wounds or scabs is just the animal kingdom doing its thing—you probably have a shrimp trying to become the alpha animal in the tank by picking on the others. (When they molt they are vulnerable, and show their wounds. The scab will disappear when the shrimp molts the next time.) More than one or two wounded, and you might have a bigger issue. If you find several, then they are fighting because they are stressed. We also look at the bellies to see if they have feed in them or not. It's important to know if they are eating well.

Any dead ones we find are thrown in the freezer to sell as bait, so I have a way to dispose of the dead shrimp, and get paid for it as well. Shrimp are popular with catfish, bass, and walleye fishermen. But now that we have such good survival rates, I do not have a lot of bait for sale anymore.

I already talked about how important our bacteria are to making an aquaculture systems work. The bacteria are not an easy thing to grow. It takes about a year to get to the right levels when starting a tank from scratch. Since we reuse our water, our bacteria levels get more stable over time. For these specialized bacteria, the ammonia from the shrimp waste is food. But just like us, the bacteria need more than one kind of food to perform at their best. They need a daily carbon boost to divide and grow at a good rate. We use simple syrup (a mix of equal parts water and sugar) as our carbon source, but you can also use molasses.

We have had the same water in our tanks for a long time. Honestly, we didn't even now that most other operations dumped their tanks and started from scratch every time. We do extra testing on our water just to make sure nothing is building up in it from the excess feed or the water we add to replace what is lost to evaporation.

For example, we test for copper and iron because our feed has some of these minerals in it. We are just making sure there is no build up from the feed that goes into suspension.

We also had our well water tested before we started using it to fill our tanks. Whether you are on well or city water, we recommend you do the same. There are lots of testing companies out there. You need to have well water tested to make sure it is safe for drinking, so please go one step further and have the water tested for pesticides. You don't want to start a shrimping operation with water that's contaminated with pesticides! Once you know what is in your water source, you can get the proper filters to remove the bad stuff before it goes into your tanks.

One thing you won't need to worry about is water hardness. The harder the water, the better for the shrimp. Remember, the ocean does not have a water softener. So NO water softeners or reverse osmosis systems. Water softeners use salts that you don't want in your tanks, and RO systems remove minerals that your shrimp need to grow and stay healthy.

If you are using city water you need to find out if it is chlorinated. If so, there are precautions you can take. Chlorinated water has to air out for at least three weeks before you put it in your shrimp tanks. I know the normal recommendation is seven days, but I don't take any chances.

Understanding the water at all phases of the operation will be key to your success. Not everyone will have the patience do this. You have to be committed to a minimum of three hours a day, seven days a week for an 8-tank system. This is not a "get rich quick" scheme. It is a baby step process. But with dedication and hard work you can get there.

Taking care of the water is how we get the survival rates we do. We average over 70% in our production tanks and over 90% in our nursery and intermediate tanks.

Before you read this book, I'll bet you didn't know that shrimp love to jump. I know I didn't, until that day one ended up in my ponytail. They will jump when they get excited or are startled. If they are close to the surface, they might get startled by a voice or loud noise and they will jump right out of the tank. Problem is, they don't always make it back into the tank and may land on the floor.

We call these shrimp "suiciders" and we sell them as bait. We cannot put them back in the tank. Even though we keep our floors very clean, even a tiny amount of bacteria from the bottom of a shoe or dust from clothing could grow and alter the bacteria in the tank. It's the same reason we sanitize our hands and tools before we touch the water. You don't accidentally want to introduce flesh-eating bacteria that destroy the shrimps' outer layer, or build a colony of bacteria that are not beneficial to the growth of the shrimp. I cannot stress it enough: we have to be very, very careful with our water.

Chapter 6

Saving Our Oceans

Most people do not realize that aquaculture is one of the fastest-growing industries in the US, and there's a reason for that. The huge amount of seafood we import from other countries is loaded with hormones and antibiotics, and may be harvested from waters contaminated with pesticides, industrial chemicals, and harmful bacteria. Nearly 96% of the shrimp we consume is imported to this country. Out of the 96% we import, only 2% gets inspected, and out of that 2%, about 60% is rejected. This is what we are consuming. The reason the majority of shrimp gets rejected is due to unacceptable antibiotic and hormone levels. These harmful antibiotics and hormones are banned for use in the United States, but are getting back into this country through imported food. The USDA is doing its best, but it simply is not good enough. They need our help.

I have always wondered why more people do not raise seafood. Our oceans are constantly being misused and over-harvested. They are literally dying. We need to help them survive by decreasing our fishing, reducing our carbon footprint, and by buying seafood from only renewable sources. Nearly half of the world's population depends on the oceans for their primary source of food. By killing the oceans, we are only killing ourselves.

A lot of research is being done on ocean-based seafood farming, which I think is wonderful. The over-harvesting, over-fishing, and polluting we are doing are turning wild-caught seafood into a bad choice.

Let's talk a little bit about how shrimp are harvested from the ocean. The most common technique uses a large net. Usually, the net is dropped to the bottom to trap the shrimp, and then pulled back up with the shrimp caught inside. The problem with this is that the fishermen bring up many different species of fish, turtles, and other marine life, along with the shrimp.

They proceed to drop everything on the deck of the boat and then sort the shrimp from the other species. Then, they throw the dead fish, turtles, eels, and anything else they accidentally caught, back into the ocean to become food for sharks and larger fish. In a sense, you are screwing up the circle of life, if you will. Some sharks no longer have to hunt for their food because it is dropped into their laps.

You are also killing off species that might not have ordinarily become food for sharks. There are news reports every day about how the ocean is dying. The coral reefs, which provide homes and food for thousands of types of fish and other ocean life, are dying as well.

If we continue along this path, we will not have anything in the ocean to eat and if the ocean dies, the planet dies. It is as simple as that. We currently have a very large population on our planet, and it continues to grow. The increase in population means an increase in the food supply. Why not grow it inland? If we have the space, the water, and the time we can do this together.

Let's give our oceans a chance to heal and save themselves. It's the only way we can save our planet.

Chapter 7

Bumps Along The Way

Earlier, I talked about the bacteria in our shrimp tanks, and why it's so important to keep it at just the right level—not too high and not too low. The bacteria form aggregates, or colonies, called biofloc, that stays suspended in the water column. The bacteria in the biofloc remove metabolic waste from our tanks. As you know by now, metabolic waste is not a good thing. Shrimp waste is high in ammonia, which is toxic. Some of the bacteria convert the ammonia to nitrite, and others convert the nitrite to nitrate. Nitrates at low levels are not harmful to your shrimp. The biofloc acts like a filter system, so we never have to exchange our water. The system cleans itself. The water in my system is over seven years old!

You could accomplish the same thing with biofilters, but they're expensive and take up space. Most facilities are already short of floor space.

As I noted earlier, the biofloc also becomes an additional feed source for the shrimp, which decreases the amount of commercial feed needed, again reducing costs.

Texas A&M University did research on recirculating systems for 20 years, but using rectangular raceway tanks rather than the round tanks we use. They had some very good results using this recirculating system as opposed to a clear water system; which is where you are constantly exchanging water as it gets dirty. Think of the fish tank at home—the filter only does so much and every so often you have to empty the tank and start over. This is very difficult, messy, and hard on your fish, right?

The system we use in our tanks is slightly different from the one Texas A&M used, but due to their many years of research on the subject, they are the best resource for information on how the biofloc works. We adapted their techniques to fit our needs and available space.

We raise our shrimp in swimming pools that are circular tanks. We chose to use the swimming pools to keep the initial cost down. Our initial setup costs were $500,000 to install the 6-tank system, the building, and the heat source needed for the water. In addition, we paid a $35,000 consultant fee.

In the beginning we used 18' x 48' round swimming pools equipped with a 1-horsepower (hp) pump placed directly in the center of the pool for recirculating the water. This was another company's design, which we used for about 2-1/2 years.

We lost about one million shrimp our first year. Some were lost due to power outages in Benton County but most were lost because of faulty or failing pumps. When it comes to something as delicate as shrimp, you only have about one hour without power before the shrimp start dying.

We lost 100% of our shrimp twice in the first year because our emergency generator failed to come on during a power outage. We had purchased a brand new generator to avoid these types of issues. Unfortunately, we didn't realize it was not functioning properly because it was not installed correctly. The auto-start feature was not working at all, so basically, our brand new generator wasn't coming on when the power went out.

The first 10 times (YES, I SAID 10 TIMES) this happened, we were on-site and were able to get the thing running manually, so we only lost about 20% of our shrimp. However, the 11th time the power went out the generator locked us out. We couldn't get it to come on manually and our tech guy was 45 minutes away.

By the time he got to us and got the generator running our dissolved oxygen levels were at 3.0, or as we call it *Death Mode*. Over the next couple of days we pulled thousands of dead shrimp out of the tanks. On day four, six tanks had dead shrimp in them. Overall, we lost over 120,000 shrimp or 3,000 pounds ($27,000 was our production cost on that many shrimp) in a single day.

They say you are not a true aquaculturist until you lose one million of whatever you're growing. Most aquaculturists never actually lose a million. We did it in the first year; we must really be pros!

Not only did we have problems with our power going out, our pumps were causing us major issues. Eventually we knew we had two choices: Either come up with another recirculation system, or get out of the shrimp farming business altogether. We were losing way too many shrimp, literally throwing money down the drain!

The pumps we originally started with were going down nearly every day. Fortunately, we were there to fix the problems most of the time and we didn't suffer too much loss from the poor quality of the pumps initially. But I really got angry when nine of our pumps went down in a one-week period. Most of the time we were at the farm and could get the pump out and fixed in less than one hour. Remember I had only six tanks at this time.

The final day we had a pump go down, it was on a tank which had 225 pounds of shrimp in it, earmarked to go to a festival that day. The truck was there to pick up the shrimp and I had 225 pounds of dead shrimp. The pump had gone down overnight, and over the next five hours, I could only get 100 pounds of shrimp from the other five tanks.

Darryl came up once the truck left and suggested various things we could try. But I was SO done with these pumps. We had already designed a system without pumps; we needed to get it up and running. In a fit of temper, I ended up slicing the pool with a knife and said "Now fix the pump." It made it very easy to clean up the dead shrimp and sell them to a bait store. Darryl decided to try using smaller pumps on the larger tanks. It still took us a couple of years before he would move the large production tanks to a pumpless system.

During our first year, we also struggled with dismal survival rates. We would order in 120,000 PLs and only 40,000 would survive in our nursery system. That was only a 33% survival rate, which we were told was fantastic. I don't know about you all, but 33% of anything isn't all that great.

Darryl and I agreed that just because it was the industry standard, and how most people were performing, losing two-thirds of our investment each month was simply not good enough for us. How were we going to survive such dismal numbers?

Chapter 8

A New Way of Doing Things

When we first got things up and running, we talked to as many experts as we could find. We took in all the advice about the correct way of raising shrimp that we could get our hands on. We emulated their nurseries, we copied their filter systems, we used their water temperatures as a guide. And we lost a million shrimp in that first year. All we found out was that none of them really knew the best way to do it, only their way to do it.

The Baby Shrimp

After seven months of what I would call failing miserably, I asked my husband if I could do the nursery *my own way*. Remember, I had no experience in the farming world other than cleaning out hog pens or moving pigs around. But still, he agreed to let me try. *How much worse could it get?*

First, I came up with my own design and my own way of doing the baby shrimp, or the nursery. We changed how we fed, how often we fed, what we fed, warmed up their water temperature, and a few other things. Darryl wasn't expecting more than a 40% survival rate.

After my first attempt at doing the PLs my way, we got a 96% survival rate. Okay, that's probably a fluke, right? No way it's ever going to happen again. The following month we got a 94% survival rate. Suddenly, we had way too many PLs and not enough production tanks. We were still ordering over 120,000 PLs per month! We now order half of that amount, and are able to supply 13 shrimp farms plus our own.

In order for us to know our survival rate, we rely on the hatchery for their initial number. Then we will literally count 100 PLs at about .5g in weight and about 1/2 inch in length three times, and calculate an average. I recalculated those numbers probably 50 times. I could not believe our survival rate was this high—it was unheard of!

In two months we went from a 33% survival rate to a 96% survival rate. Amazing, right? We had to figure out what to do with all the extra PLs. As it turned out, there were a couple of other shrimp farms in Indiana, including the one we bought our system from, and they needed PLs. We had extra. So, we started a whole new branch of our business—selling PLs to other shrimp farmers to jump start their systems. Who knew this would become part of our business? We certainly didn't.

Since changing everything we were taught in March of 2011, not once have we fallen below a 90% survival rate for the PLs. Things were finally looking up.

The Intermediate System

After about a year we were ready to expand our facility. The plan was to add six more tanks, but on a smaller scale, as this was all the space we had. The barn we expanded into had housed our combine. Darryl had an idea—he thought we should consider using the new facility as an intermediate system. We had over a 90% survival rate in our nursery division, but then after transferring the PLs into the production tanks, the survival rate was plummeting to only about 40%. That made absolutely no sense to us.

Why not let the PLs grow a bit more before putting them into the production tanks? You see, I thought the pumps in the production tanks were what was killing the baby shrimp. They seemed too small and delicate for the movement of the pumps which maintain the suspension system—keeping the bacteria suspended in the water. If the bacteria settle on the bottom of the tank, the water becomes anaerobic, killing the shrimp.

Our intermediate tanks did not use pumps, so I thought this would make for a smoother transition from the nursery. Then we could keep all six production tanks as production tanks, with shrimp only being in there for three months. It moved them out faster.

However, we still had the problem of keeping the water in suspension without a pump. Darryl came up with the idea of the "volcano" and airlift system. We had no idea if it would work but it did. We were actually getting survival rates from the intermediate tanks that were just as high as from the nursery.

And, since this design was working, we implemented it in the production tanks as well. Granted, it took me a couple of years to convince Darryl this was the way to go. He thought that pumps were necessary to keep everything in suspension in 4000 gallons of water.

The Volcano Effect

Now that we got rid of all the 1-hp pumps, we are able to run on air only. Air is drawn in from the outside through a 1- or 2-hp regenerative blower. The more tanks you have, the bigger the blower or the larger the number of blowers needed. The regenerative blower just pushes the air through a 4" PVC pipe and down into the tanks. We use airlifts to push the oxygen into the tanks.

The airlifts are PVC pipes with air stones in them to push water, causing two separate currents in the tank. By pushing water up through the PVC pipe, we are causing more oxygen to go into the water. We no longer need pumps to put oxygen into the tanks.

I mentioned what we call the "volcano effect" for the center of the tanks. It is a solid drum with an airline or air stones around the center of the drum. The drum is solid so no water can get inside to cause water quality issues. This drags the waste and excess feed off the bottom of the tank, runs it up the side of the solid drum, and then out on top of the water. Hence a "volcano effect."

The biofloc must be in **suspension** at all times. This is crucial to the survival of the shrimp. If it settles out, it will cause "sludge" at the bottom of the tank and the shrimp will suffocate. In the sludge there is no oxygen (it is anaerobic) and that causes further water quality issues.

Heating and Facility Arrangement

We heat our water using a hydronic radiant heat system. One of the reasons we like using this type of system is the fact that you choose which fuel source you use for heat. With most aquaculture systems you have only one choice. Radiant heating systems can run off any number of energy sources—solar, geothermal or wind energy; natural gas, LP, or propane; and even wood or pellet stoves. We use LP gas because it is easiest for us.

In a hydronic system, warm water is transferred from a water heater into a closed PEX tubing loop that sits directly inside the tanks. The heat radiates from the PEX tubing into our tank water. We started with a residential water heater, but they're just not built to serve a large system. We now use tankless commercial water heaters.

This is more heat efficient and energy efficient. All I am heating is the water. Not the air, not wasted space. Heat from the water will heat your air. You never want to heat the air to heat the tanks. It would be so hot it would be very hard to breath and you wouldn't want to work in the facility. If you can't work in the facility you will start losing shrimp.

Even if your temperature outside is 105 degrees F., your heat source may still come on. The shrimp have to stay at a constant temperature. There is very little room for variance in the water temperature. As I've mentioned before, 1-2 degrees either way can be a killer.

Whatever heating system you will be using, it does not run 24/7. It is on demand. It only turns on when necessary and shuts off when the desired water temperature is reached.

We have three barns running and have built a fourth one to expand into. In our original building we have seven production tanks—one 18' swimming pool, two 16' swimming pool and four 12' fiberglass tanks. The intermediate barn has seven 14' swimming pools. Our nursery barn houses five nursery tanks (4' x 8' x 2') and two crawfish nursery tanks. If we could do it over again we would line up all the tanks in one facility.

Our newest barn is 5000 sq. feet (50' x 100'). This barn houses 12 14' swimming pools, 12 crawfish tanks (4' x 2') and a salt water aquaponics system. We have 3' alleyways in all our buildings to make getting around easier. We also leave 1.5 feet around each tank. This gives you space for the settling tank and space to walk around the tank. Trust me; we learned this from a big mistake we made in the beginning. We could not walk around our tanks, and if a shrimp gets caught and dies between tanks, it will stink!

The buildings must be very well insulated. We use 1" of spray foam, R39 fiberglass insulation, a vapor barrier and over all this we use a product called Duraplate. Duraplate is semitrailer siding. It's manufactured in Lafayette, Indiana, about 20 minutes from us, and we get it at their scratch and dent sales. We chose to use it because of ease of cleaning. You are raising food and you want something that is easy to clean. You can use any type of product that is easy to clean, like barn siding, for example.

The ceilings are only 8 – 8-1/2' high. That probably seems low but there is little point to heating a bunch of extra air. You pay to heat, so don't waste money heating space you don't need to. The ceilings have 26" of R40 blown insulation with a vapor barrier. We used barn siding on the ceiling.

Drainage

We also recommend drainage for your building. We use trench drains down the center of the buildings. The floor is on a slight slope. Again, this is for ease of cleaning. We do not have to mop or squeegee our floors. We just wash them about five times a day due to the amount of foot traffic coming and going through the farm.

Power Usage

We use a very minimal amount of electricity. We run one and two horsepower regenerative blowers and we have lights over each of the tanks. Keeping it simple is the premise of our system. We try not to overthink it. We are just trying to maintain Mother Nature at her best inside a tank with no interference from external elements. This means we can raise a much cleaner and healthier product with fewer water quality problems.

Chapter 9

The Standards We Set

I would never have thought shrimp farming was for me, but I think you've figured out by now that I love everything about it. I am the first one to work in the morning and usually the last one to leave. I love the research we are doing and continue to do. I have a goal to make Indiana the shrimp capital of the world, along with a 96% survival rate for every farm.

Why not become the shrimp capital of the world? Indiana is located in one of the best places to offer fresh shrimp throughout the country—right smack dab in the middle!

Shrimp farming isn't easy, but anything worth doing or having isn't easy. It makes you appreciate it more.

We do have high standards and we urge all shrimp farmers to set high standards for themselves. Right now, there are no government regulations for us. If we continue doing it the way we are doing it now, we may never have regulations. But it only takes one person to cut corners and ruin it for everyone else.

We are asked all the time why we help other people—our "competition," if you will. I help because I cannot do this by myself. We cannot be an island and hope to grow.

As an island, I have borders and can only go so far. If I work with other shrimp farmers I feel that my borders are endless and I can go anywhere.

Everyone asks about competition. So far, we are just a small industry trying to educate people on the hazards of the shrimp they are eating from the grocery store. We are still just a drop in the ocean. If someone put a shrimp farm across the road from me I would be the happiest person in the world. We would bring in more traffic and we would be able to help each other. Why are there three gas stations on most corners? More traffic to the area. Not one does better than the other. It is pretty equal.

Chapter 10

So, You Want to be a Shrimp Farmer?

What it takes to build and outfit a shrimp facility is not as complex as you might think. You can use any type of building as long as it is well insulated, has a concrete floor, and a built-in drain.

Having a well-insulated building lets you do this anywhere. We have helped set farms up in so many types of buildings. We have converted chicken barns, hog barns, turkey barns, school gymnasiums, new buildings, garages (not my favorite), and even the basement of a manufacturing company.

You don't need a specific type of building other than it needs to be well insulated so you don't heat what you don't need to heat. You should have concrete floors for easy cleaning. Putting a system on a dirt floor is not really a good idea. I personally wouldn't want to walk around in mud all day. Drains are a must. We recommend trench drains that go down the center and length of your building to make for simple and easy cleanup. You can use "can drains" but then you have to squeegee the water to the drain. Using trench drains will make your life easier.

You will have to have your incoming water source tested. You can use either well or city water, but there are different protocols for each type of water. If you use well water, it needs to be safe for drinking and free of pesticides. You need to be 100% sure of what is in your water before you begin filling your shrimp tanks.

Testing will also help you decide what type of filter you will need for your water, if any. You will need to pay close attention to the levels of metals or minerals in the water. High iron, for example, is devastating to shrimp. The water at our farm is very high in iron, so we use a "household" brine filter to remove it before the water goes into the shrimp tanks. Hardness is a good thing. Actually, the harder the water the better it is for the shrimp. That's because calcium and magnesium are critical for the normal functioning of their little hearts, muscles, and nervous systems. Calcium also plays a big role in the molting process and how well their new shells harden.

If your facility will use city water, you have to worry about the chlorine content of the water. There aren't any filter systems that remove 100% of the chlorine from the water. The only way is to oxygenate it out of the water by letting it sit for a minimum of seven days. You will have to have two spare vertical water holding tanks. These tanks will be aerated with air stones. One tank must always sit for at least seven days before you use the water to refill your shrimp tanks as they lose water to spillage and evaporation.

You never want to add water straight from a tap into the tank. When you add water straight from the tap it is 54 degrees F and the tank water is 86 degrees F. The shrimp do not like change in their water temperature and that will cool down the tank too much.

You will need to install 115-volt single phase electrical service to run your equipment. We run 1.5 to 2 hp regenerative blowers to pull the air from the outside and into the tanks. You will need lights overhead.

Light fixtures can be very minimal. We recommend lights over the tanks and then a set of lights down the alleyway. The lights down the alleyways will be left on 24/7. Shrimp DO NOT like the dark. If you put them in total darkness you startle them and they will jump out of the tanks. When you turn the lights back on, they will start jumping again. Remember —stressed shrimp do not grow very well.

Backups are very important. Shrimp can only survive one hour without power. They may not all die at one time, but in four days the entire tank will be dead, due to the stress of no oxygen in the water. You will need an auto-start generator. It must be an auto-start, not the type that you start with a tractor. We have one of those for the house but it sometimes takes over an hour to get the tractor started in order to run the generator. So make sure you have an auto-start generator. I also recommend the extra service that most companies offer of texting you when your generator goes on and off or needs any maintenance.

You can never have enough backup. We recommend along with the generator that you also have an auto dialer that will call you when the power goes out for more than seven minutes. Generators do not always work as they should and if you get a phone call telling you are without power, you know the generator is not working. This hopefully leaves you with enough time to get to the farm and start the generator.

If you are more than 20 minutes away, have a neighbor you can call to come and start your generator for you. We actually have two neighbors we can call if we are away from our farm.

I know this sounds like a lot of caution, but as you know if you've read this far, we have lost 100% of our shrimp twice due to the generator not starting, and a third time we lost 70% of our shrimp. Backup for backups only makes sense. It also gives you a bit of peace of mind if you live off the farm.

We recommend hydronic radiant heat for heating the tanks. It's essentially the same as a heated tile floor. We run the heating system directly inside the tanks so it is energy and heat efficient. Heating the water will heat your building and not the other way around. If you were to try to get the air warm enough to heat the water, it would be much too hot for you to work in your barn.

What is nice about our heating system is its flexibility when it comes to energy sources. As I mentioned previously, some aquaculture systems have heaters that run on only one energy source. With ours you can use what is best for you in your area.

If you live in a very warm climate I am still going to recommend a heat source as a backup. Shrimp do not like temperature fluctuations. Heaven forbid you have an unseasonably cold night in southern Florida without a backup heat system. You can't afford it; all your shrimp will be dead in about four days. A heat source doesn't have to run unless you need it. We actually shut down our heat supply during the summer months. It may run five days a week during the winter months and then maybe once a week during the summer months.

We talked a bit about drainage earlier. You are working with water, and you will have spillage from dipping into your tanks and moving shrimp between tanks. You want to be able to easily clean your floors, so having built-in drainage is a must. This is why I recommend the "trench drains" in the middle of your alleyway to make cleaning a breeze. I don't know about you, but I don't want to clean all day. Using the trench drains saves you time and energy when it comes to cleanup.

You will need to keep your facilities clean. Even though most states do not have any regulations for shrimp farming (yet), you are still selling a food product. You do not want your consumer to get sick because a shrimp lands on a dirty floor and you then sell it. It not only hurts your business but it hurts mine as well.

Time management is one of the most important pieces for success in this business. You must be committed to seven days a week and a minimum of three hours a day. You cannot work one or two hours a day, or skip days. Three hours is the minimum amount of time it will take to test nine tanks every day and work on feed schedules, tank cleaning and floor cleaning. When you need to move or sell shrimp, this will take even more time as you will be harvesting by hand using a net. It is not as intimidating as you may think. We have 19 production tanks, seven intermediate tanks, 10 nursery tanks, 20 crawfish tanks, two tilapia tanks and a salt water aquaponics system that we test every day. There are only two of us that do the testing. I do all the nursery tanks and a staff member does all the rest. She starts at 8:00 AM and has usually finished testing by noon. We have one person who gets the feed ready and does a 9:00am feeding of 100g of feed till we know how much total feed will be needed for the day based on the water quality. If you add more tanks or add a nursery system to your operation you will need to add more time. Trying to care for an eight tank system in less than three hours will not work. You won't be paying enough attention to the water quality. I can't stress enough that this is the most important part of this process. I learned through our arrogant stage that skipping a day or two can cost you a tank full of shrimp.

Chapter 11

Regulations and Certifications

Regulations

For now, aquaculture falls under agriculture in the federal rules and regulations. In the state of Indiana, there are no rules or regulations as long as we sell our shrimp live. Selling them live guarantees the freshness of the shrimp and we bypass the health department.

If we were to kill the shrimp, that would be considered processing, and as such, we would then be bound by the state's health department regulations. Basically, we would be subject to state inspections of our facility at least twice a year and we would be required to have food server licenses and a HACCP (hazard analysis and critical control points) plan in place.

Right now, the USDA does not inspect our facility because we are selling the shrimp live. Essentially, it is no different than if I was selling a live hog or cow off my property. However, once you put a knife to livestock, you are considered a processor, which will subject you to the rules and inspections of the USDA.

Even if you only refrigerate or freeze your shrimp for consumption, this is also classified as a form of processing and will put you under the rules of your state's health department.

We highly recommend you check with your county extension office, your health department, and the economic development director in your county to see if there are any rules or regulations you may have to adhere to. In some states you have to register with the Department of Natural Resources (DNR) just so they know who is raising what.

Even though we are not being inspected by any agency, we have decided to adhere to all the rules and regulations and then some, as if we were being inspected. We choose to do it this way so our customers know we go above and beyond in order to give them a quality product they cannot get anywhere else.

The other reason I do this is due to the "cantaloupe incident" a few years ago. Some of you may remember that there was an Indiana cantaloupe farm that had to issue a recall due to a salmonella outbreak at its facilities in Indiana and several other states. I threw away my cantaloupe and never bought another cantaloupe the rest of that year.

Some farmer took shortcuts and caused a problem not just for himself but also other farmers. I always think about that, because I know I was not the only person who wouldn't buy a cantaloupe that year and I'm sure the industry suffered. So if we all stay above board and work very hard to keep our facilities clean, and get certified even though we are not asked to do so, it puts us ahead of the game.

If you cut corners and don't take care of your water, or keep your facilities clean, you are not only hurting your own business, but you are hurting mine as well.

Certifications

As I mentioned before, there are no rules or regulations for us to follow unless there is a county ordinance in your area. Be sure to double check this as a mistake here could be costly for you. We also understand that some counties do not know what to do with you because "shrimp farming" does not fall under any category. It is still very vague.

There are some certifications you can get that help you gain credibility with your county as well as your customers. We recommend looking into whether or not your state has a livestock certification. Indiana does, but it took us a while to get ours. We had to get some of the rules changed in order for what we do to fit into its parameters. Shrimp are livestock, but most of the regulations pertained to hogs, cows, chickens, ducks and turkeys.

Whether your state has a livestock certification or not, we recommend obtaining a seafood HACCP certification as well as a food server license. Both of these can be done online, followed by a one-day class. It is all worth it in the end because not only will your customers appreciate the extra effort you went through to offer them a quality product, it also shows your county and state that you want to do this the correct way.

From left: Former Indiana Secretary of Agriculture Ted McKinney, Karlanea Brown, Russell Brown, and Kimmi DeVaney, former Agritourism and Livestock Programs Manager, Indiana State Department of Agriculture.

Chapter 12

Tours and Community Involvement

Tours are great way to make some money in the first year when you are not making a lot of money on your shrimp. "Agritourism" (or "agrotourism") is one of the fastest growing industries in the country right now.

Agritourism, as it is defined most broadly, involves any agriculturally based operation or activity that brings visitors to a farm or ranch.

Agritourism makes up about 20% of our business. We have groups come to our facility as well as individuals. They do take up some of your time (about an hour per tour in our case) but you can charge admission. It was actually our county economic development director that suggested it to us. We initially had no intention of doing tours but what better way to get paid for advertising? Visitors don't want to know all the scientific stuff that goes into shrimp farming, they just want to know your story. You need to be able to answer questions like:

Why did you get into shrimp farming?

What type of system do you use?

What is so special about your system?

What makes your product better?

Why would I want to buy your shrimp?

The added bonus is how few people have actually seen a shrimp up close, still alive, and being raised indoors. They don't even have shrimp in aquariums. At lease I have never seen one. Maybe those ghost shrimp, but that's not what we raise.

You can do mini festivals. Pair up with a brewery or winery and a chef and you have a great little festival for people to come and see what you are about. People visit your facility and they tell their friends, who then tell their friends. There is no better form of advertising than word-of-mouth. One leads to two, two leads to four, and so on. You get the idea. We do not need to advertise; our business is word of mouth.

It's very important to be involved in your community and be in the public eye. Education is a huge part of what we do. You have to remember, the average shrimp eater currently walks through the frozen section of the local grocery store and picks up a bag or box of frozen shrimp. It is our job to educate the public as to why what we do is better, and why our product is better.

Chapter 13

Do You STILL Want To Be A Shrimp Farmer?

Can you talk about shrimp? I mean REALLY talk about shrimp? All day, every day, talk about shrimp? You cannot be embarrassed by the "Forrest Gump" association. You have to be willing to talk about your product so people understand why they want it. You have to be able to tell them why it's so much better than the frozen stuff they get in the grocery or big box stores.

It isn't easy work but it isn't impossible, either. The bottom line is in the DETAILS. It is paying attention to each and every detail, each and every day. It's about noticing that the water looks different today; it's about noticing that your shrimp seem to have less energy than they did yesterday; it's about noticing the baby shrimp are bouncing around too much. And it's about testing, testing, and more testing.

My mother-in-law used to say, "You know, we can read all the books on shrimp farming, the very few that are out there anyway. But in the end, it doesn't matter, because the shrimp have not read any of the books." They are going to do what they do. They don't care what I want them to do. Remember it is about the water. As long as you take care of the water, the water will take care of the shrimp.

I believe we need more shrimp farmers and other types of aquaculture farmers, but it takes a special type of person. This job is part farmer, parent, teacher, naturalist, conservationist, politician, and a little bit of activist all rolled into one. How many jobs can you say that about?

If you think you can handle all of these roles, and are ready to become a shrimp farmer, ask yourself these questions:

- Can you commit to seven days a week? Shrimp cannot take care of themselves and the water can change very quickly on you. I am not saying you cannot take a vacation. You just need to have a person you trust to take care of the testing while you are gone. Remember, I am on the road two weeks out of every month and I have two people that I leave behind that I trust to make sure I have a shrimp farm to come home to. And live shrimp!
- Can you commit to at least three hours a day? Some days more than three hours? Doing three hours a day for a start up business is doable. You have to keep your day job for now, so if you can work three extra hours a day you can start to build your business. Anything less than three hours a day and you won't be around long. Your survival rates will be low and you will not make the money you were hoping to make. Without shrimp, you make no money.
- Do you have about $100,000.00 to get started with an 8-tank system? This will cover everything you need minus the building. We cannot factor in the building. Your building can be anything that works for you as long as it is well insulated, has drains in the floor, and has the space for a lab, mechanical room, office, sales room and feed storage area. The best case would be zero windows in the production area or greenhouses, as they affect your ability to completely control the climate, which makes them not ideal. We have set up shrimp farms in all kinds of buildings—chicken and turkey barns, hog barns, school gymnasiums, storage facilities, and even brand-new buildings.
- Are you willing to work in an 82-degree, 98% humidity environment? The barns are warm but comfortable if insulated correctly. Are you willing to wear shorts and t-shirts all year long? If you cannot work in the humidity, or on some days, a sauna-like atmosphere, this is not

the business for you. I tell everyone I have one of the world's most expensive saunas.

- Are you willing to do the testing every day? Seven days a week? ***There is no exception to this rule.*** Maintaining the water quality is like guarding the keys to your bank vault. Doing the testing is how you know what is going on inside your tanks. I will be honest, the testing gets boring after a while, so you will have an "arrogant" stage and be tempted to skip testing. Remember what happened to me when I didn't test for a few days? Testing everyday will help keep your shrimp alive. Something worth having is worth working for.

Darryl and Diamond, our farm dog, answering emails at the end of a long day of harvesting shrimp.

Chapter 14

Let's Get Shrimping!

Okay, so now you know all about what it takes to be a shrimp farmer. Are you still interested or did I scare you away?

Here's the last thing that I can offer you that I did not have: I can help you get up and running. I can help you start farming the world's greatest seafood and you can do it ANYWHERE.

We offer a system that is simple and easy to manage as long as you take care of the water. We have always believed that if something is hard to learn and hard to do, most people will likely not continue. You will not be making money the first year, or the second year. Perhaps you will make money the third year, IF your survival rates are high enough. Why would you want to work at something that is this hard to work in and this hard to learn at the same time?

This is why we start you off with a smaller tank system that will produce about 80 pounds of shrimp at a time with an 80% survival rate. It is not a lot of shrimp but you will be able to build up the bacteria you need and get a better survival rate as time goes on. Like I said earlier, this is not a get rich quick scheme and you will have to work to get the numbers you want.

When you start getting more and more poundage of shrimp out of a tank you will feel like all the work was worth it. Remember, you are starting a business that you only need to commit to for about three hours a day for the first year. As you start to expand and add tanks, you will need to commit more hours.

So if you are ready, we would love to be your advisors and help you get your shrimp farm started successfully. We offer consulting, with a step by step manual and an 8-tank system that will include your tanks, the volcano and airlift system along with the complete settling tank. We also include the air lines and valves that lead from your air source.

We can also help you set up a commercial-ready system if you can commit to eight hours a day with two employees from the start.

Please contact me at rdmshrimp1@gmail.com and let's get shrimping.

In your email, please share your location (city/state/country), the number of tanks you want to start with, and if you have an estimated launch time for your shrimp farm.

You can also find more information on our website at RDMShrimp.com.

We will respond in four days with information, pricing and available times for a one-hour phone consultation. If desired, a private tour of our facility can be arranged. The private tour is $200 and lasts about four hours. It covers everything we do. We sit down at the end and answer any remaining questions you may have.

So, let's get Shrimping!

Darryl & Karlanea Brown

Owners

RDM Aquaculture LLC

rdmshrimp1@gmail.com

(765) 583-0052

RDMShrimp.com

Testimonials

Prof. Kwamen Quagrainie, Purdue University, Agricultural Economics
Author of *Profitability in Shrimp Farming*

With growth in the local foods industry and more people wanting to know the source of their food, indoor production of Pacific White shrimp (Litopenaeus vannamei) has shown to be a promising investment for farm families in the Midwest. RDM Aquaculture in Fowler, Indiana, is a pioneer in this effort, utilizing farm buildings and other farm facilities for shrimp production. Consumers are recognizing the quality of local shrimp and are paying premium prices. Shrimp producers are motivated by the fact that they could produce and market fresh (never frozen) high-quality shrimp products to compete with imported frozen shrimp. With the right investment strategies, indoor shrimp production could become a major Midwestern agricultural industry.

Jeff H., Missouri

I was taught in business school that most successful businesses are unique, have little competition, and have high barriers to entry. After spending 14 years in the corporate world and deciding a change was needed, I found that the shrimp farming industry was an intriguing business opportunity. I knew I had found what I was looking for when I visited RDM Aquaculture with my brother, James. In addition to many entrepreneurial advantages, shrimp farming has many social benefits such as supplying customers with a clean protein source and being environmentally friendly. I am relatively new to the shrimp farming world but I can already see why this industry has seen consistent growth.

Brenda and Randy T., Ohio

We were looking for a way to diversify our funding interests. Shrimp farming has allowed our family to do so. In a small space we are able to provide a quality product to a growing market.

About the Author

Karlanea and Darryl Brown are the husband and wife team behind RDM Aquaculture. Together, they run one of the most successful salt-water shrimp farms in the United States, all from their hometown of Fowler, Indiana.

Karlanea is the VP of Operations and the pioneer of doing things her own way. She wasn't happy with the low survival rates of her shrimp and had the drive to change the way things were done. She revolutionized indoor shrimp farming. Since implementing their "Volcano Effect" system, the couple has seen incredible success and a dramatic increase in their survival rates.

Karlanea is in charge of keeping the shrimp happy, healthy and full of life. And water testing as well. She is also in charge of helping her fellow shrimp farmers be better at what they do. She is such an advocate for what she does that she has even lost her voice once or twice talking about shrimp.

RDM Aquaculture received the Indiana Innovation Award in 2015 for its tank system, running on air only. The farm is Indiana Livestock Certified and is a member of Indiana Grown, where new generations of farmers, foodies, and families have new ways of growing their business, creating jobs, and eating healthy.

Karlanea loves what she does. Despite the fact that she holds a degree in fashion design, and spent years in interior decorating, she wouldn't change a single thing. She's the happiest she's ever been farming shrimp and plans to still be doing it when she is 90 years old. She is thinking of retiring when she turns 100 years old, to do some traveling.

Made in the USA
Monee, IL
04 November 2026

67379702R00039